MIX
Papier aus verantwortungsvollen Quellen
Paper from responsible sources
FSC® C105338

Adegbite Adeleke Adesipo

Basic Studies on the Dual-Polymer Flocculation of Iron Hydroxide

Anchor Academic
Publishing

Adesipo, Adegbite Adeleke: Basic Studies on the Dual-Polymer Flocculation of Iron Hydroxide, Hamburg, Anchor Academic Publishing 2017

Buch-ISBN: 978-3-96067-199-2
PDF-eBook-ISBN: 978-3-96067-699-7
Druck/Herstellung: Anchor Academic Publishing, Hamburg, 2017

Bibliografische Information der Deutschen Nationalbibliothek:
Die Deutsche Nationalbibliothek verzeichnet diese Publikation in der Deutschen Nationalbibliografie; detaillierte bibliografische Daten sind im Internet über http://dnb.d-nb.de abrufbar.

Bibliographical Information of the German National Library:
The German National Library lists this publication in the German National Bibliography. Detailed bibliographic data can be found at: http://dnb.d-nb.de

All rights reserved. This publication may not be reproduced, stored in a retrieval system or transmitted, in any form or by any means, electronic, mechanical, photocopying, recording or otherwise, without the prior permission of the publishers.

Das Werk einschließlich aller seiner Teile ist urheberrechtlich geschützt. Jede Verwertung außerhalb der Grenzen des Urheberrechtsgesetzes ist ohne Zustimmung des Verlages unzulässig und strafbar. Dies gilt insbesondere für Vervielfältigungen, Übersetzungen, Mikroverfilmungen und die Einspeicherung und Bearbeitung in elektronischen Systemen.

Die Wiedergabe von Gebrauchsnamen, Handelsnamen, Warenbezeichnungen usw. in diesem Werk berechtigt auch ohne besondere Kennzeichnung nicht zu der Annahme, dass solche Namen im Sinne der Warenzeichen- und Markenschutz-Gesetzgebung als frei zu betrachten wären und daher von jedermann benutzt werden dürften.

Die Informationen in diesem Werk wurden mit Sorgfalt erarbeitet. Dennoch können Fehler nicht vollständig ausgeschlossen werden und die Diplomica Verlag GmbH, die Autoren oder Übersetzer übernehmen keine juristische Verantwortung oder irgendeine Haftung für evtl. verbliebene fehlerhafte Angaben und deren Folgen.

Alle Rechte vorbehalten

© Anchor Academic Publishing, Imprint der Diplomica Verlag GmbH
Hermannstal 119k, 22119 Hamburg
http://www.diplomica-verlag.de, Hamburg 2017
Printed in Germany

Abstract

The efficiency of synthetic flocculants in flocculation process has been identified by several authors over the years, and this makes it applicable in the industry. Nowadays, studies are now focused on the combination of polymers for optimal performances. In this study, the physicochemical conditioning of ferric hydroxide suspension (obtained sludge from the lignite mined sites in the Lusatia region of Germany) was investigated with polymers of different properties as single polymer conditioning (C492 and A130) and as dual-polymer conditioning (C492+N300, A130+N300 and C492+A130) using Jar test. C492 is a cationic flocculants, A130 is an anionic flocculants, and N300 is a non-ionic flocculants. Turbidity of the supernatant and the sludge volume index (SVI) were the considered assessing parameters under a dosage range of 0.4, 0.8, 1.2, 2.0, 2.8 and 3.6 mg/gTS. The obtained optimization results was compared to the actual separation with respect to specific filtration resistance from pressure filtration. The result shows that C492+N300 has no significant positive effect compare to C492 for the physicochemical optimization while A130+N300 produced a lower SVI but more turbid supernatant compare to A130. Based on the obtained values and observation of the types of flocs formed during the flocculation processes, it could be deduced that polymer bridging predominates with C492 and N300 while charge neutralization predominates with A130 flocculants. And based on their actual separation process, C492+N300 has the lowest specific filtration resistance during the actual separation which indicates better filterability. And in most cases, the physicochemical optimization result does not correlates with the actual separation result. However, in general, the higher the dosage, the more compacted the flocs structure generated and the lower the filtration resistance. Also, worthy of note as could be evident from this study is that the timing for pressure filtration is not a determinant for its specific filtration resistance. However, in a haze, choice of dual-polymer flocculation should be based on the properties of the flocculants, appropriate dosing concentration and the desired assessing parameters.

Keywords: solid-liquid separation, flocculation, synthetic flocculants, dual-polymer, pressure filtration

Table of Contents

Abstract i
List of Figures iv
List of Tables v
1. Introduction 1
   1.1 Study Objectives 2
2. Literature Review 4
   2.1 Solid Liquid Separation 4
   2.2 Theory of Flocculation 5
      2.2.1 Flocculation and Energy Barrier 5
      2.2.2 Flocculants 6
      2.2.3 Flocculation Mechanisms 7
      2.2.4 Factors Affecting Flocculation 10
   2.3 Flocculation Assessment 10
   2.4 Charge Characterization 10
   2.5 Filtration Technique 11
      2.5.1 Pressure Filtration 12
      2.5.2 Specific Filtration Resistance 13
      2.5.3 Tendency of Deviation 14
3. Materials and Methods 15
   3.1 Materials 15
   3.2 Method 15
      3.2.1 Charge Quantification 15
      3.2.2 Flocculation Tests (Jar Test) 17
      3.2.3 Pressure Filtration 18

4. Result and Discussion 20
   4.1 Physicochemical Optimization Process Results 20
      4.1.1 Optimum Dosage for Single Polymer Conditioning 20
      4.1.2 Optimum Dosage for Dual Polymer Conditioning............................ 21
   4.2 Comparison of the Physicochemical Results 23
   4.3 Pressure Filtration Process 24
      4.3.1 Single Polymer Filtration Result 24
      4.3.2 Dual polymer Filtration Result 25
   4.4 Comparison of the Specific Filtration Resistance 26
5. Conclusions and Further Studies 28
References 29
Appendix 33

List of Figures

Figure 1: Solid-Liquid separation process 4
Figure 2: A) Potential energy representation. B) The electrical charge of particles 6
Figure 3: The graphical representation of flocculation 8
Figure 4: Formation of floc 9
Figure 5: Curves showing the various tendency of deviation from the normal 14
Figure 6: Working principles of particle charge detector (Product sheet manual) 16
Figure 7: Set-up for Jar-Test experiment using Flocculator 17
Figure 8: Schematic diagram of the pressure filtration apparatus 19
Figure 9: Graphical representation of the physiochemical process for C492 and A130 21
Figure 10: Dual polymer flocculation for C492+N300, A130+N300 and C492+A130 22
Figure 11: Specific Filtration Resistance for C492 and A130 25
Figure 12: Specific Resistance value for C492+N300, A130+N300 and C492+A130 ... 25

List of Tables

Table 1: The "best" Floc Characteristics for different separation processes 11
Table 2: Typical dosage range for dewatering process of biosolids. 12
Table 3: Properties of the different flocculants employed in this study 15
Table 4: The physicochemical pre-treatment result 20
Table 5: Parameters for calculating the specific filtration resistance 24
Table 6: The specific filtration resistance for all the flocculants considered 24
Table 7: Specific filtration resistance at optimum dosage 26
Table 8: Correlation between SFR and assessed parameters 27

1. Introduction

As the world population increases, demand for potable water increases, likewise the generation of waste. This reveals that sludge in general; either as a result of mining or as effluent from the industry, household or other sources requires proper treatment in order to attain expected standard for water reuse and groundwater recharge. In solid-liquid separation process; its efficiency and the quality of the separated liquid (supernatant) is usually of interest considering the involved economic and environmental consequences. Brostow, et al., (2009) noted that of several separation processes; gravitation, coagulation, flocculation, etc. flocculation is the fastest. In the treatment of suspensions and fine particle systems; it plays a significant role, and in solid-liquid separation process; it remains indispensable for both upstream and downstream (Oyegbile, et al., 2015).

Research has proven that solid-liquid separation can be improved by the use of synthetic flocculants in flocculation process. It increases the effective particle size by forming larger flocs and breaks down the suspension stability followed by the release of the liquid phase. It is an efficient and cost-effective strategy and has become widely employed in several industrial applications (Fan, et al., 2000). This is because it generates larger flocs which can sediment and be separated easily. Polymers used in flocculation can destabilize these particles either by bridging, charge neutralization, depletion flocculation or electrostatic patch (Chaiwong & Nuntiya, 2008). Ebeling, et al., (2005) pointed out some advantages of the use of polymers over alum, ferric chloride and other aids that have been in use before. Few of these includes its low dosage requirement, reduction of sludge production, easier storage and mixing, no pH adjustment requirement, its effectiveness for smaller particles and improvement of the floc resistance to shear forces.

Over the last few decades, research has been focusing on the combination of polymers for optimal performance. Several research works have discussed the benefits of using two-component flocculants as compared to single component. Britt, (1973) was one of the first groups of researchers to work on this topic and has been further investigated in several other studies. Razali, et al., (2012) stated that dual flocculation provides a better control of flocculation kinetics than single-flocculants as well as improved flocs structure. In addition to that, research by Wu & Theo, (2009) on poly (ethylene oxide) and carboxylate phenolic resin confirm that dual flocculants improve the richness of flocculation but, it depends on the concentration of the two components used. They noted that dual component flocculants may consist of two polyelectrolytes, two polymers, or polyelectrolytes and a nano-colloid. They further stated that one of the components adsorbs on the surface of the particles while the other bridges these polymer-coated particles. Use of dual polymer can produce a synergetic effect and complexation for flocculation process (Fan, et al., 2000); an example is an increase in the bonding effect. In a similar work, Yu & Somasundaran, (1993) investigated on alumina flocculation with polystyrene sulfonate and cationic polyacrylamide polymers which confirms that the flocculation result obtained from two polymers premixed was much lower than when the polymers were used individually.

1.1 Study Objectives

The general aim of this study is to compare the single and dual-polymer conditioning of ferric hydroxide suspension with respect to the subsequent solid-liquid separation efficiency by pressure filtration. The sludge is part of the waste product obtained from the lignite mining areas in the Lusatia region of Germany). And the specific study objectives include;

- ➢ Physicochemical treatment of ferric hydroxide suspension using flocculants with different charge densities.
- ➢ Assessment of the physicochemical optimization process with respect to the turbidity of the supernatant and the sludge volume index (SVI).
- ➢ Determination of optimum flocculants dosage with respect to the measured parameters and separation efficiency.

This will be achieved by the performance of a bench scale flocculation and dewatering experiments in order to obtain the optimum dose. The experimental tasks involve charge quantification of ferric hydroxide suspension using the particle charge detector (PCD), determination of the charge densities of the flocculants that were selected using

PCD, determination of the optimum dose of single polymer followed by dual polymer based on the assessment parameters

Turbidity and sludge volume index were used among other assessing parameters because the turbidity explains how much of the particle has been removed from the supernatant and it is one of the most popular used performance criteria. Likewise, the sludge volume index is an evidence of the sedimentation efficiency and the dewatering of the flocs formed. The specific filtration resistance to be obtained from the pressure filtration experiment is an important parameter in determining the dewatering behavior of the conditioned sludge, it explains the filterability of the flocs formed. Ferric hydroxide is an example of mineral sludge from the mining industry and can as well be from wastewater treatment plants, few studies have so far been done on it.

2. Literature Review

2.1 Solid Liquid Separation

Wakemann, (2005) established that solid-liquid separation generally aims at separating two phases from a suspension with four main targets (1) recovery of valuable solid component, (2) recovery of the liquid, (3) recovery of both phases (solid and liquid) and (4) recovery of neither phases. Therefore, in order to ensure a convincing efficiency in solid-liquid separation, the techniques and design of any system to be employed must be able to handle these four main separation processes described in Figure 1 below.

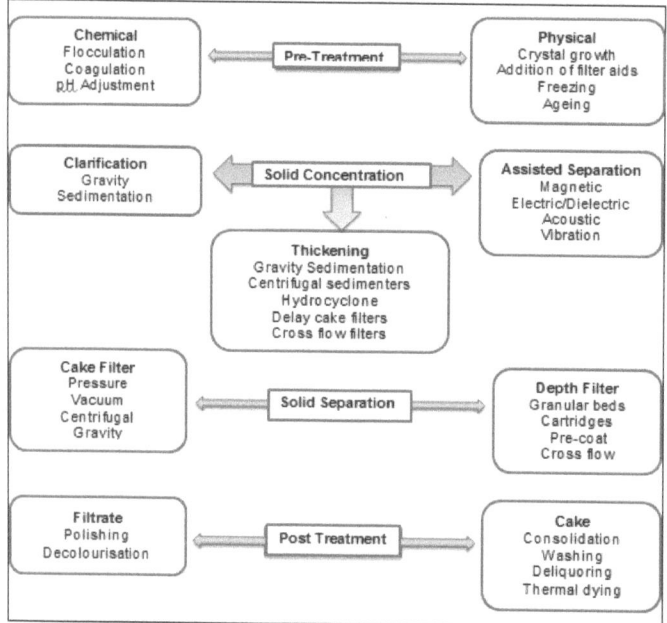

Figure 1: Solid-Liquid separation process. Adapted from Wakemann, (2005)

The main advantage of pre-treatment is to enhance the efficiency of subsequent separation processes, it reduces the water binding forces between the solid particles. During the pre-treatment process, changes may occur in the suspended liquid properties (particularly chemical process) (Wakeman & Tarleton, 2005). An appropriate and targeted manipulation of the sludge properties will, therefore, improve the downstream thickening or dewatering process. The flocculation experiment in this study enhances the agglomeration of the particles thereby building larger flocs which were easy to be separated in the solid separation stage in which pressure filtration was employed. Factors that influence the dewatering abilities of sludge include the concentration, organic matter content, and the colloidal nature of the sludge. Wakemann, (2005) suggested the recognition of some technical alternatives for the concentration of solid which includes; magnetic, electrical or sonic force field which he termed "assisted separation" techniques by separation technologist. Post-treatment processes involves steps to improve the quality of either the solid or the liquid. This depends on the aim of the solid-liquid separation based on various standards to be met.

2.2 Theory of Flocculation

Flocculation process has been discussed by several authors and needs to be understood in order to harness it appropriately. The theory of flocculation in this section focuses on the reasons why flocculation is needed (energy barrier) followed by the mechanisms of flocculation with its efficacy and the factors that affect flocculation.

2.2.1 Flocculation and Energy Barrier

Most particles in suspension exhibit a surface charge which may be as a result of ionization of surface groups, ionic substitution, and uneven distribution of constituent ions or due to specific adsorption of ions (Moody & Norman, 2005). Although settling under gravity is known to be the simplest means of separation but, fine particles of 10μm do not sediment in an economically within a reasonable amount of time (Brostow, et al., 2009). This is as a result of a build-up of potential energy which causes the force of repulsion as a result of similarly charged particles thereby hindering sedimentation if it predominates unlike the force of attraction known as London-van der Waals forces which helps the particles to come together (Figure 2a). Moss & Dymond, (2013) explains the two main sources of this repulsion force (1) adsorption of water onto the surface from the surrounding which forms solvation layers (2) as a result of electrical charge of the particles on their surfaces (Figure 2b). In general, aqueous suspension within the pH of 4 and above usually carry negative charge while in acid (of pH less than 4), positive charged suspension is common.

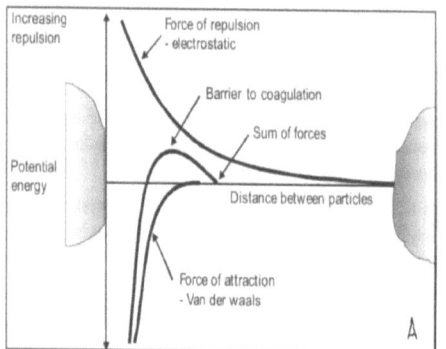 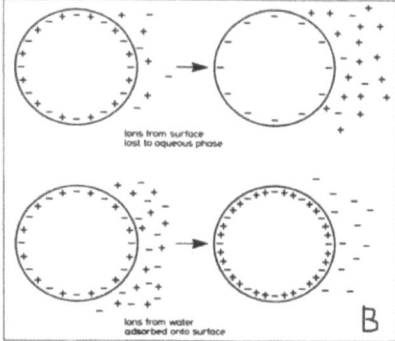

Figure 2: A) Potential energy representation. B) The electrical charge of particles (Moody & Norman, 2005)

Moss & Dymond, (2013) pointed out that charge repulsion effect is very much noticed with small particles because of the greater surface area to mass and charge to mass ratio. This barrier as a result of the repulsion force can therefore be overcome by mechanical agitation, reduction of electrostatic charge (which includes addition of inorganic salts, adjustment of pH and by the addition of inorganic or organic coagulants), and by physical bridging between the particles which involves the use of high molecular weight polymeric flocculants (Moody & Norman, 2005).

2.2.2 Flocculants

Flocculants are in two categories; organic and inorganic flocculants. Salts of multivalent metals like Iron and Aluminum are the inorganic flocculants mostly applied at a higher concentration but, organic flocculants are polymeric in nature (Brostow, et al., 2009). Inorganic flocculants usually produce a very large amount of sludge and are affected strongly by pH variability. Organic flocculants, on the other hand, is of two types; natural polymers obtained from natural sources like starch, alginates, glues etc. and synthetic polymers. Although natural flocculants may be biodegradable and environmentally friendly, they have several disadvantages like higher dosage requirement, loss of floc strength and instability in the solution. Based on the monomeric units charges of the functional groups, together with the molecular weights, synthetic polymers are classified into cationic, anionic or non-ionic, which may either be of low molecular weight, medium or high molecular weight. Most of the flocculants are based on polyacrylamide (PAM), poly ethylene oxide (PEO) and its derivatives. Poly-acrylamide is non-ionic and the variation in its ionic properties is based on copolymerization with other monomers. Copolymerization of acrylic acid with acrylamide or partial hydrolysis of polyacrylamide is the basis for preparing anionic polyacrylamides while copolymerization of acrylamide with quaternary ammonium derivatives of acrylamide forms the basis for preparing cationic polyacrylamides. Development of polyacrylamide-based flocculants nowadays will likely depend on 'molecular architecture' concept which will enhance the solid-liquid separation process (Hande, et al., 2014). Of recent, there is a new generation of flocculants that is being produced in addition to conventional flocculants with the aim of increasing the efficiency of the flocculation process called the 'unique molecular architecture' (UMA) (Pearse, et al., 2001).

The difference in the flocculants charges help provide a means for adsorption to particles surface through electrostatic attraction and also cause the extension or uncoil the polymer molecule with a linear-like structure in order to accommodate more particles. Hande, et al., (2014) and Yan, et al., (2004) stated the factors that influence the adsorption characteristics of polymer for flocculation of fine particles which include; the type of the polymer, polymer molecular weight and chain length, conformation, charge density (degree of ionization) and the functional group. Anionic flocculants seems to be more available commercially because of its cost and its higher molecular weight as compare to cationic flocculants because the more the molecular weight of flocculants, the more efficient is its use for flocculation.

2.2.3 Flocculation Mechanisms

According to a report by Q'Max Solution Inc. on 'Flocculation process in drilling fluids' as well as Moss & Dymond, (2013), there are two main basic important actions in particle flocculation (1) adsorption of the polymer unto the particles and (2) aggregates formation (Figure 3), and it is in in support flocculation mechanisms described by several authors.

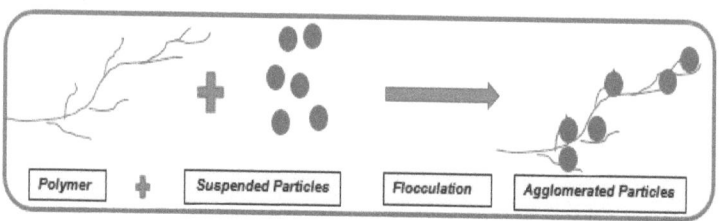

Figure 3: The graphical representation of flocculation

As it was earlier stated that most suspension with pH of 4 and above mostly contain particles that are negatively charged; cationic flocculants will then be expected to be the best flocculants for flocculation, this is only possible in the case of charge neutralization and particle attraction of polymer but, it is not for the bridging mechanism of flocculation. When the surface and the reagent are of opposite charge, adsorption occurs which results in non-specific electrostatic interaction. And theoretically, positively charged mineral ought to adsorb anionic (negative) and vice versa, but having a large amount of polymers adsorbed; further adsorption is likely to cease because recharging and stabilization of particles should occur and the particles will then be of opposite charge compared to the initial charge (Moss & Dymond, 2013).

The bridging mechanism is the basis of aggregation of particle in situation where nonionic or high molecular weight and low charge polymers are used with particles of similar charge in which segments of the same polymer molecule attach to other particles. But, the complex situation arises when polyelectrolytes (with low, medium or high molecular weight) are used together with oppositely charged particles whereby any of these three forms of mechanism may occur, it includes charge neutralization, electrostatic patch and flocculation bridging. Polymer bridging and charge neutralization are the common mechanisms of flocculation as described by several authors. In charge neutralization, there is the presence of particle surface charge opposite to the ionic polymer functional groups. And bonding can occur through hydrogen bonding, electrostatic and hydrophobic interactions, chemical bonding, ion bonding, Van der Waals forces and by the formation of insoluble salts (Somasundaran & Das, 1998; Mpofu, et al., 2003; Hogg, 2000; Gregory & Barany, 2011).

Although the bridging mechanism has been refined over time, but the main basis remains unchanged. The bridging mechanism occurs when the polymer chain first adsorbs partly onto a surface while the loop and tails protrude into the solution and get attached to a second surface (Brostow, et al., 2009). Several hypotheses exist for flocs formation mechanism, the first is the zeta potential reduction by adsorbed charge molecule thereby making the particles to form floc by van der Waals force attraction. The second hypothesis is the attachment of the polymer simultaneously to two particles and more adsorption occurs as polymer molecule is contracted as described in Figure 4a.

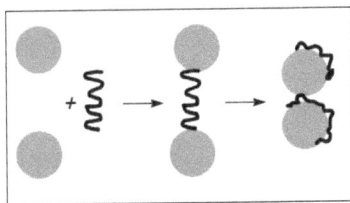 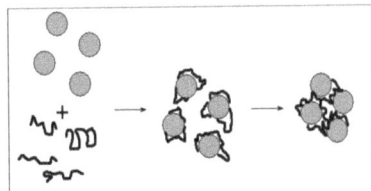

Figure 4: Formation of floc (a) Simultaneous attachment of polymer to two particles by its two ends (b) Attachment of the polymer at different points with loops attaching to other particles (Moss & Dymond, 2013)

The third hypothesis assumes that at different points on the particle surface, the polymer molecule will be adsorbed having loops which vary in length and can project out of the surface. The collision of these partially covered particles thereby leads to the bridging effect, this is shown in Figure 4b. Based on this, one may conclude that adsorption of the polymer molecule on particles can happen before the formation of the bridge.

Strongly adsorption ability of the polymer is expected for bridging of polymer and either high molecular weight cationic or anionic can be used. Higher molecular weights polymers can in general help improve bridging flocculation and for an effective by the bridging flocculants, there ought to be an enough small distance between the particles for the loops and tails to connect two particles (Ebeling, et al., 2005). However, it has been noticed that at over-dosage of the polymer, the whole particle can be coated with polymer leaving no space for bridging to take place with other particles which are referred to as "hair-ball effect" (Ebeling, et al., 2005).

2.2.4 Factors Affecting Flocculation

There are two main factors that flocculation rate is dependent upon; the collision frequency and the collision efficiency (Mclaughlin, 2010). This collision is affected by the mixing energy and time, the pH, temperature, quantity of particles and the type and dosage of flocculants used. Flocculation efficiency is usually affected if the optimum dosage is altered. This is the exact amount of polymer that is needed for optimum flocculation result is related directly to the multi-particle adsorption amount. Beside this, there is a directly proportional relationship existing between the optimum polymer or solid ratio and the surface area of the solid. A decrease or increase in one alter the demand for the other. The type of dosage in terms of the molecular weight is another important factor. It has been observed that there is a tendency for each molecule of the polymer to adsorb onto a single particle if a lower molecular weight polymer is used which limits flocculation efficiency on further addition of polymer. However, there is a greater adsorption if a higher molecular weight polymer of the same type is used.

2.3 Flocculation Assessment

There are two common methods of flocculation efficacy evaluation which are jar test and settling test (Brostow, et al., 2009). Although some believe that settling test is more efficient because jar test results sometimes reveal minima and/or maxima, apart from that, it is time-consuming and suspension-specific. The nature of the suspension is a better determinant of the best method. For this study; jar test was used following already established experimental procedure by previous studies in the field (AWWA, 2011; Freese, et al., 2003).

2.4 Charge Characterization

Characterization of sludge is very important because it provides information about the state of the sludge and its likely behavior during treatment and disposal. Of the several parameters that can be used in the characterization of sludge (dry residue, loss of ignition, sludge volume index, compressibility, state of charge, Capillary Suction Time CST, etc.), charge state measurement is important due to its influence in conditioning. This helps in chemical saving by controlling the flocculants dosages, increase the dry content of sludge through dewatering and optimized the selection of flocculants. One means of achieving this is the streaming potential measurement using PCD03 (Particle Charge Detector). Counter ions separated by dissociated molecules or particles is measurable as streaming potential (mV: millivolts). The sample reaches the equivalence point when all the charges are neutralized (streaming current of 0 mV). The measured potential sign is an indicator of the charge in the sample. The potential depends on the electrical conductivity of the sample dispersion, sample viscosity, molecular weight and particle sizes of the sample, dimension of the measuring cell temperature and cleaning of the measuring cell.

2.5 Filtration Technique

Several Solid-liquid separation techniques have been discussed by several authors. Lee, et al., (2005) listed some solid-liquid separation techniques that can be employed based on the floc characteristics (Table 1).

Table 1: The "best" Floc Characteristics for different separation processes (Lee, et al., 2005)

Solid-liquid separation	Best floc characteristics
Clarification	Large floc, loose floc structure, high permeability, with charge reversed flocs
Sedimentation	Large floc size, dense floc structure, regular floc shape, compressible, with no surface charge
Filtration	Large floc size, loose floc structure, high floc strength, high permeability, incompressible, with surface charge.
Centrifugation	Large floc size, dense floc structure, incompressible cake, high floc strength
Flotation	Low floc-density, high floc structure, uniform floc size, hydrophobic floc surface
Consolidation	High elasticity, dense floc structure, low bound water content, no surface charge
Electroosmosis	Low compensation voltage, with surface charge, large floc size, loose structure, high permeability

There are typical range of dosage that can be used for dewatering as described by several authors, but, few works have been done so far on ferric hydroxide which limits the availability of its data. Table 2 below shows the typical dosage range for dewatering of biosolids according to (Girovich, 1996)

Table 2: Typical dosage range for dewatering process of biosolids (Girovich, 1996).

Application	Typical Polymer Dosage gram dry polymer/kilogram dry solids
Belt filter press	3-6
Centrifuge	2-10
Vacuum filter	5-10
Pressure filter	2-3
Vacuum assisted drying beds	15-25

2.5.1 Pressure Filtration

Based on the generation of the required pressure drop across filter medium, there are three different classification of surface filters in surface filtration (unlike depth filtration), it includes; vacuum, centrifugal and pressure (Svarovsky, 2000). Applying a suction on the filtrate side of the medium is the basis of the driving force for vacuum filters while a rotating perforated basket which is fitted with a filter medium are the constituents of a centrifugal filter. The centrifugal field action is of two main effects; pressure drop is created across the filter medium as a result of the centrifugal head of the suspension layer that rotates in the basket and the liquid being pulled out of the cake and the medium.

But for pressure filters filtration, the liquid pressure which is developed by pumping or by the force of gas pressure in the feed vessel is the basis of the driving force. Important to note is the ability to generate pressure drop across the medium in excess of 1bar which makes it preferable to the theoretical limit of vacuum filter. In addition to that, pressure filter yield higher output coupled with drier cakes and better filtrate clarity (although not always). Svarovsky, (2000) noted that an increase in pressure drop for compressible solids causes a decrease in the permeability of the cake thereby reducing the rate of filtration in relation to the pressure drop.

Several models to describe the physical process involved in filtration have been developed like cross-flow filtration, deep bed or depth filtration, blocking filtration etc. but, this study focuses on cake filtration. Here it is assumed that subsequent filtration takes place on the cake after the formation of the first layer of cake during filtration whereby, the filter medium which acts as the main impermeable barrier at the beginning, thus serves as a support. At a constant flow rate (dV/dt), there will be a proportional linear increase in the pressure drop to the solid deposited quantity.

Pressure filtration can be done either at constant pressure or at a constant rate in which pressure drop across the medium is kept constant in the case of constant pressure or keeping the filtration rate constant by adequate adjustment of the pressure in the case of a constant rate (Ghosh, 2006). The target of most research study is to determine how efficient and fast filtration experiment can be but, the following factors have to be considered as presented by Darcy's filtration equation (equation 1)

$$Q = K\left(\frac{\Delta P A}{\mu R}\right) \tag{1}$$

Where Q is the flow rate (through the differential of volume against time), K is the permeability, A is the area while ΔP is the pressure drop across the filter medium, μ is the viscosity and R are the medium resistance. The main factors include the thickness of filter cake, area of the filter medium and the pressure drop.

2.5.2 Specific Filtration Resistance

Specific filtration resistance has become one of the most important parameters that can be used for determining the drainage or dewatering behavior of conditioned and unconditioned sludge. It explains the filterability of the flocs to be filtered (i.e. Ferric hydroxide in this experiment). It describes how difficult it is to filtrate, unlike permeability which describes the ease of the liquid to pass across. In this study (at constant pressure), adapting the calculations derived by Wakemann, (2005) and Svarovsky, (2000), the slope (m) of a linear graph equation y=mx+c (equation of a straight line) between t/V (s/m^3) against V (m^3) is used to evaluate the specific filtration resistance (α) and the intercept (c) is used for calculating the medium resistance R (this is not considered in this study). Because the cake properties are not of interest in this study and the cake thickness is not fixed, equation 2 below can be used to determine the average specific resistance ($α_{av}$).

$$\alpha_{av} = 2A^2 \Delta P \times \frac{slope}{C\mu} \tag{2}$$

Where A is the filtration area (m²), ΔP is the pressure difference, slope (s/m⁶) is the value (m) gotten from the graph of t/V (s/m³) against V (m³), μ is the viscosity (N.s/m²) and C is the solid concentration (kg/m³).

2.5.3 Tendency of Deviation

There is a tendency for the curve of the graph t/V (s/m³) against V (m³) to deviate from the normal curve as explained by Ripperger, et al., (2012). Figure 5 below shows the various possible curve that could be obtained.

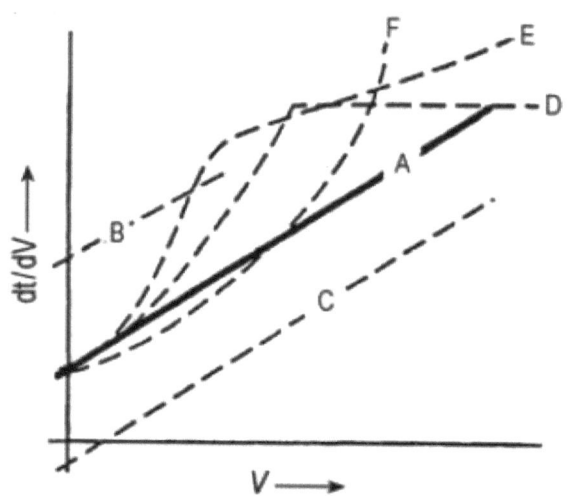

Figure 5: Curves showing the various tendency of deviation from the normal (Ripperger, et al., 2012)

(A) is the normal theoretical linear curve, (B) some solids have settled before filtration and apparently increase the resistance, (C) the starting point was not measured correctly, filtration has started before timing begins, (D) the solid settles completely; increases the cake growth speed (E) settling of only coarse particles whereby the remaining fine particles are filtered and (F) trickling of fine particles through the cake thereby blocking the pores (cake/media). Other cases might be because of the incorrect beginning and endpoint of filtration, or the unusual properties of the suspension which may not be considered like bubbles or oil droplets, non-Newtonian rheology of the liquid and the isometric, etc.

3. Materials and Methods

3.1 Materials

The average result of the concentration of Ferric hydroxide determined using the oven at 105^0C for 24 hours shows its concentration as 1.18% and it falls within a neutral pH range (7.60 – 7.65). It has a charge density of 192.119 ueq/g which was quantified together with other flocculants using the particle charge detector (PCD). The particle charge detector is a product of BTG MÜTEK GMBH with the working principle described below. It is used in combination with an automatic titrator. Table 3 below states the flocculant types, the molecular weight, the type of charge, and the charge densities.

Floculateur 11198 (Figure 7 below) used for the jar test is a product from Bioblock Scientific having the highest speed of 200rpm which was the speed used for fast mix. The turbidity meter that was used; Turbiquant 3000 IR is a registered trademark of Merck KGaA, Darmstadt, Germany. And the filter paper (598/1) used as the filter medium has a diameter of 45mm and size range between 12-25µm.

Table 3: Properties of the different flocculants employed in this study

Flocculants	Molecular weight	Type of Charge	Titrated with (0.001N)	Charge Density
C492 (cationic)	High	Low	Pes-Na	266.8 (ueq/g)
A130 (Anionic)	Low	Medium	Poly-DADMAC	1036 (ueq/g)
N300 (Non-ionic)	High	No charge	Poly-DADMAC	35.2 (ueq/g)

3.2 Method

3.2.1 Charge Quantification

Due to the form of flocculants that was used (solid form), stock solution (1.0 g/l) was prepared in order to save the preparation time and conserve the flocculants. This was later diluted to appropriate concentration. Each of the solutions was used within a period of 5 days for effectiveness and accuracy of result. For the quantification of each flocculants, 10ml of the sample was poured into the measuring cell of the PCD03, the piston ring was appropriately positioned inside the measuring cell which was fixed to the PCD meter. The motor was switched on and allowed to run for about 1-2 minutes for the stabilization of the potential. The burette tip of the automatic titrator was inserted for the titration. Poly-DADMAC polyelectrolyte was used for A130 and N300 while Pes-Na was used for C492. Ferric hydroxide was quantified using Poly-DADMAC polyelectrolyte.

3.2.1.1 Operating Principle of PCD03

The Charge Particle Detector meter works according to the principle of streaming current detection. There was a gap which was defined between the piston and the wall of the vessel (measuring cell). The piston moves through a motor that drives it up and down instituting a fluid flow whereby free counter ions are swept away and separated from the adsorbed sample in the cell. A current is then induced by the counter ions in the built-in gold electrode which displays the streaming potential (Figure 6).

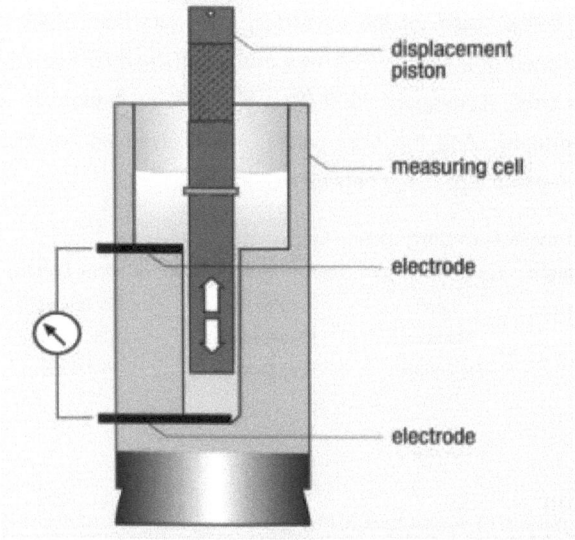

Figure 6: Working principles of particle charge detector (Product sheet manual)

3.2.1.2 Polyelectrolyte Titration

For quantitative charge measurement, polyelectrolyte or colloid, titration was conducted with the Particle Charge Detector (PCD) which uses the streaming current to identify the point of zero charge 0 mV (equivalence point). An oppositely charged polyelectrolyte was used with an already known charge density and added to the sample as titrant. The titrant charges then neutralize the existing charge of the sample to the isoelectric equivalence point which was the termination point for the titration. The common polyelectrolyte used include Pes Na 0.001N (Anionic polyelectrolyte) and poly-DADMAC 0.001N (cationic polyelectrolyte). The specific charge quantity q (eq/g) can be evaluated using equation 3.

$$q = \frac{VC}{wt} \qquad (3)$$

Where V is the consumed titrant volume (ml), C is the titrant concentration (eq/l) and wt is the solids of the sample or its active substance (g).

3.2.2 Flocculation Tests (Jar Test)

The experimental set up is shown in Figure 7. The five beakers (representing each of the dosage range) were filled with 250ml of the Ferric Hydroxide and was placed under each paddle of the flocculator ensuring free movement to enhance proper mixing and prevent breaking of the glass. The fast mix was at 200 rpm which lasted for 1 minute while the slow mix was at 30 rpm for 5 minutes. The fast mix helps to properly mix the suspension with the polymer while the slow mix is meant for the aggregation of flocs. Careful attention needs to be given to the mixing velocity especially the 'slow mix' because the optimum size of the larger flocs formed during the slow mix cannot be reserved again when torn.

Figure 7: Set-up for Jar-Test experiment using Flocculator

The dosage range considered includes; 0.4, 0.8, 1.2, 2.0, 2.8, and 3.6 mg/gTS. The physicochemical treatment of the ferric hydroxide was performed for single polymer conditioning using C492 (cationic flocculants) and A130 (Anionic flocculants) individually. The dual-polymer conditioning was performed with a combination of non-ionic flocculants N300 with each of the flocculants used as single polymer (C492+N300 and A130+N300) at a ratio of 1:3 while the two main flocculants were as well combined together (C492+A130) at a ratio of 1:1.

The evaluating parameters; turbidity of the supernatant and sediment height was taken and recorded after 30 minutes for each of the experiment. The supernatant was taken at an average height of 3cm for the turbidity measurement. Based on the sediment height, the sludge volume index (SVI) was calculated using the formula in equation 4 below.

$$SVI\left[\frac{ml}{g}\right] = \frac{Vs\left[\frac{ml}{l}\right]}{TS\left[\frac{g}{l}\right]} \tag{4}$$

Vs is the volume of the sediment height (ml/l) and TS is total solid (g/l) of ferric hydroxide. The result of each experiments was recorded and presented accordingly in chapter 4.

3.2.3 Pressure Filtration

The set up for the pressure filtration is shown in Figure 8. The pressure filtration was performed for all the dosages considered for the physicochemical process. The pressure filter has a vertical filter medium with filtration proceeding top-down. The pressure applied through the air supply was manually regulated and being monitored using the Manometer. The filter chamber is made of metal which prevents escape of air and stabilize the pressure. The measuring scale was connected to a PC which records the mass of the filtrate per unit time concurrently. Both the supernatant and the flocs from the flocculation experiment were poured into the filter chamber after which the air pressure is applied. At lower pressure, the filtration experiment last longer and vice versa. In each of the experiment, the automatic drop in the pressure coupled with the sound produced as a result of only air coming out of the filter chamber in place of filtrate were the indicators that the filtration has been completed.

The average specific filtration resistance ($α_{av}$) was determined using equation 5 below.

$$αav = 2A^2 ΔP \times \frac{slope}{Cμ} \quad (5)$$

Where A is the filter area (m^2), $ΔP$ is the filtration pressure (N/m^2), the slope (s/m^6) is from the graph of time/volume (s/m^3) against the volume (m^3), μ is the viscosity ($N.s/m^2$) and C is the solid concentration (kg/m^3).

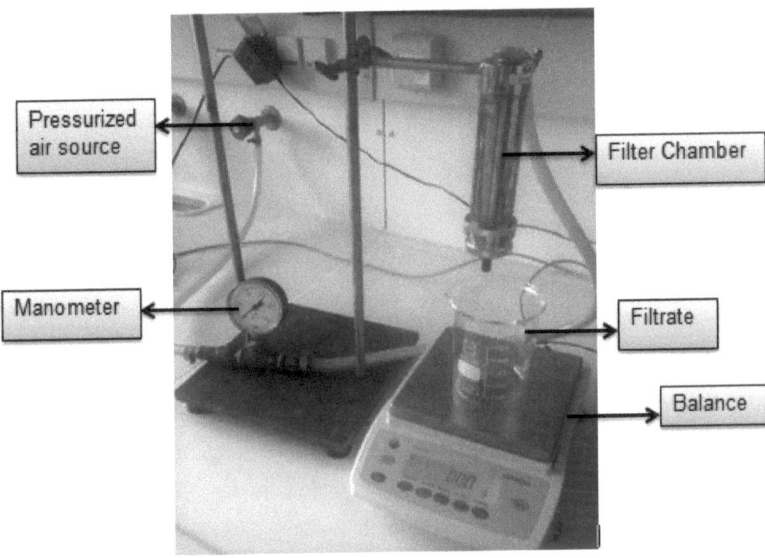

Figure 8: Schematic diagram of the pressure filtration apparatus

4. Result and Discussion

4.1 Physicochemical Optimization Process Results

The results of the physicochemical optimization process for all the considered flocculants combinations (i.e. both the single and dual) are presented in Table 4 below.

Table 4: The physicochemical pre-treatment result

Dosage (mg/gTS)	C492		C492+N300		A130		A130+N300		C492+A130	
	SVI (ml/g)	NTU	SVI (ml/g)	NTU	SVI (ml/g)	NTU	SVI (ml/g)	NTU	SVI (ml/g)	NTU
0.4	12.203	5.50	13.559	10.53	30.170	6.19	22.712	6.36	24.407	3.79
0.8	12.712	4.12	13.220	7.59	29.153	3.47	22.712	3.86	24.068	3.60
1.2	13.220	3.70	12.542	6.4	28.475	1.54	23.051	2.77	24.068	2.80
2.0	11.525	5.32	11.864	4.91	28.475	1.19	24.407	2.95	21.695	4.41
2.8	11.186	4.47	12.203	5.95	29.153	1.82	24.407	3.12	22.712	4.58
3.6	13.559	5.39	14.237	6.75	29.153	3.34	24.746	3.36	21.695	5.48

4.1.1 Optimum Dosage for Single Polymer Conditioning

Figure 9 presents the graphical representation of the physicochemical optimization results for both the cationic C492 and the anionic A130 flocculants respectively. The SVI values obtained for C492 ranges from 11.186 ml/g – 13.559 ml/g while that of turbidity ranges from 3.70 – 5.50. The optimum dosage for SVI (11.186 ml/g) was obtained at 2.8 mg/gTS and 1.2 mg/gTS for turbidity (3.70). The obtained values for both SVI and NTU below and above its optimum dosage is expected to follow a significant trend, but at 2.8 mg/gTS where optimum dosage for SVI was obtained, there is a change in the trend of NTU as well. However, the type of floc generated with the cationic flocculants C492 is large and well compacted (please see the appendix).

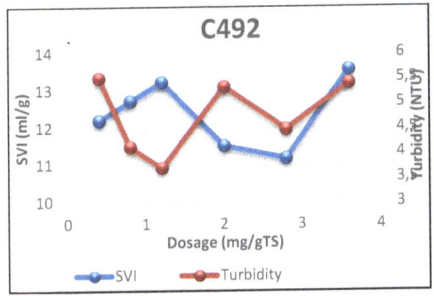 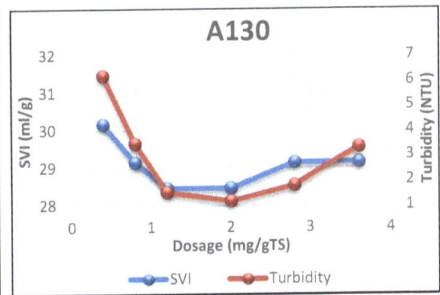

Figure 9: Graphical representation of the physiochemical process for C492 and A130

On the other hand, the anionic flocculants A130 has its optimum dosage for both SVI and turbidity at 1.2 mg/gTS (28.475 ml/g) and 2.0 mg/gTS (1.19) respectively. The result follows a regular optimization trend as compared to cationic flocculants C492, but its range of obtained values for SVI is higher (28.475 – 30.170 ml/g), while its turbidity values (1.19 – 6.19) are lower. The generated flocs with the anionic flocculants A130 are not large, and are weak (please see the appendix), however, worthy of note is its less turbid supernatant.

In relation to the properties of the flocculants as presented in Table 3 above, the result shows that with the cationinc flocculants C492; polymer bridging is involved, it has high molecular weight with low charge, and the floc as well are large and strong. But with the anionic flocculants A130; charge neutralization is involved. This is due to its properties (low molecular weight and medium charge), and the appearance of the type of flocs generated which are small and weak.

4.1.2 Optimum Dosage for Dual Polymer Conditioning

The dual polymer conditioning results obtained for C492+N300, A130+N300 and C492+A130 is presented in Figure 10. C492+N300 and A130+N300 were combined at ratio 1:3 while C492+A130 was at 1:1.

C494+N300: The obtained values of dual-polymer conditioning with C492 and N300 follows a significant regular trend. . The optimum dosage is 2.0 mg/gTS for both SVI (11.864 ml/g) and the turbidity (4.91). Range of its SVI obtained values is from 11.864 – 14.237 ml/g while that of turbidity is from 4.91 – 10.53. The visual observation of the flocs shows a bulky and large type of flocs (see the appendix).

A130+N300: The trend of the dual-polymer conditioning with A130+N300 shows that; the lower the dosage, the lower the SVI. The optimum dosage is 0.4 mg/gTS for SVI (22.712 ml/g) and 1.2 mg/gTS for turbidity (2.77). The range of obtained values for its SVI is from 22.712 – 24.746 ml/g while that of turbidity is from 2.77 – 6.36. The sludge conditioning of A130+N300 has small floc, (similar to its single polymer conditioning), and the lower the dosage the smaller the floc.

C492+A130: With C492+N300, optimum dosage of 2.0 mg/gTS (21.695 ml/g) and 1.2 mg/gTS (2.80) was obtained for both the SVI and turbidity dual-polymer conditioning respectively. The optimization follows a regular trend (especially turbidity), there is a noticeable change with SVI trend at 3.6 mg/gTS. The SVI range of values is from 21.695 – 24.407 ml/g while that of turbidity ranges from 2.80 – 6.48.

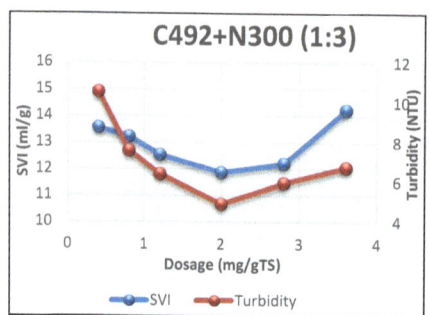
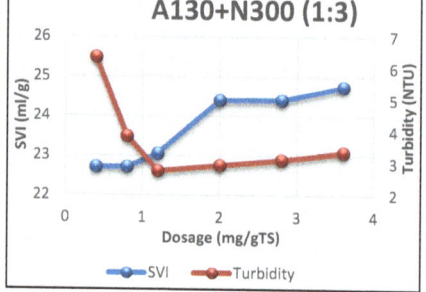
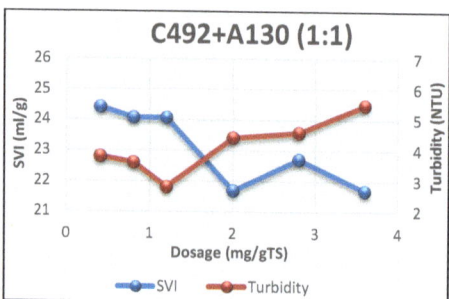

Figure 10: Dual polymer flocculation for C492+N300, A130+N300 and C492+A130

4.2 Comparison of the Physicochemical Results

The results show that there are differences in the influence of sludge conditioning process with single polymer compared to dual-polymer. Combination of cationic flocculants (C492) with non-ionic flocculants (N300) at ratio 1:3 within the considered dosage range has no positive influence on both the SVI and turbidity. This suggests being due to the similarity in their high molecular weight and low to no charge properties. On the other hand, the dual-polymer conditioning with anionic and non-ionic (A130+N300) flocculants produced a lower SVI, but more turbid supernatant. This suggests being as a result of the differences in the properties of the flocculants; A130 has low molecular weight with medium charge while N300 has high molecular weight with no charge. However, assessing parameters with C492 and A130 lies in-between.

Considering the result based on the different flocculation efficiency assessment; it could be deduced that the influence of the flocculants as a single polymer is as well reflected in its combination as dual-polymer. And it suggests that an appropriate dosage concentration and the assessing criteria is the determinant for the efficiency of dual polymer flocculation, and this agrees with Wu & Theo, (2009).

Also, the physicochemical optimization result and the visual observation of the floc structure suggests that polymer bridging mechanism predominates with C492 and N300 flocculants while charge neutralization predominates for A130 flocculants (Lee et al, 2005). This is as well supported by Ou-Yang & Santore, (1999) that charge neutralization mechanism performs over a broad range molecular weights while bridging mechanism depends highly on the molecular weight. This agrees with established facts by several authors that polymer bridging mechanism predominates with high molecular weight and low charge flocculants (like C492) while charge neutralization mechanism predominates with flocculants that has a high charge and low molecular weight (like A130).

4.3 Pressure Filtration Process

The determination of the specific filtration resistance which explains the dewatering behavior of the conditioned sludge in full scale operation is the basis of the laboratory pressure filtration experiment. Its obtained values describe the filterability of the type of floc formed from the physicochemical conditioning process. Table 5 below presents the key parameters used in calculating the specific filtration resistance and Table 6 presents the specific filtration resistance values for all the flocculants combinations considered.

Table 5: Parameters for calculating the specific filtration resistance

Parameters	Values	
Filter Area (A)	45 mm (Diameter)	1.6×10^{-3} m^2
Filtration Pressure (ΔP)	Between 0.65 – 1.5 bar	$0.65 \times 10^5 - 1.5 \times 10^5$ N/m^2
Viscosity (μ)	2.0 mPa.s	0.002 Ns/m^2
Concentration (C)	11.8 g/l	11.8 kg/m^3
Slope	From the graph	s/m^6

Table 6: The specific filtration resistance for all the flocculants considered

Dosages (mg/gTS)	Specific Filtration Resistance (m/kg)				
	C492	C492+N300	C492+A130	A130+N300	A130
0.4	2.8204×10^{11}	2.6034×10^{11}	2.1695×10^{11}	1.7356×10^{11}	1.9256×10^{11}
0.8	1.2692×10^{11}	1.3017×10^{11}	1.9526×10^{11}	1.3017×10^{11}	2.1695×10^{11}
1.2	9.8712×10^{10}	5.2068×10^{10}	2.1695×10^{11}	1.5187×10^{11}	2.1695×10^{11}
2	1.1281×10^{11}	2.6034×10^{10}	1.5187×10^{11}	1.7356×10^{11}	4.3390×10^{11}
2.8	5.2068×10^{10}	7.8102×10^{9}	1.3017×10^{11}	6.5085×10^{10}	1.6271×10^{11}
3.6	2.6034×10^{10}	1.8224×10^{10}	9.7628×10^{10}	3.2548×10^{10}	6.5085×10^{10}

4.3.1 Single Polymer Filtration Result

Figure 11 shows the graphical representation of the specific filtration resistance result obtained for C492 and A130. The two flocculants shows a decreasing trend at higher dosages. However, A130 has a noticeable change in its specific filtration resistance trend at 2.0 mg/gTS.

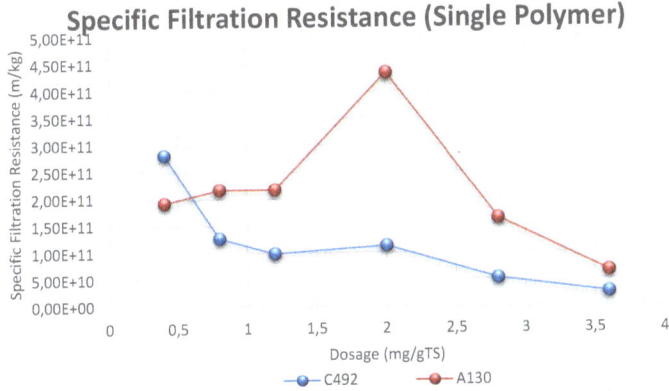

Figure 11: Specific Filtration Resistance for C492 and A130

4.3.2 Dual polymer Filtration Result

The graphical representation of the specific filtration resistance result for the dual polymer experiments; C492+N300, A130+N300 and C492+A130 is presented in Figure 12 below. Only C492+N300 does not have a noticeable change in the trend of its curve with the least resistance values as a result of its low curve.

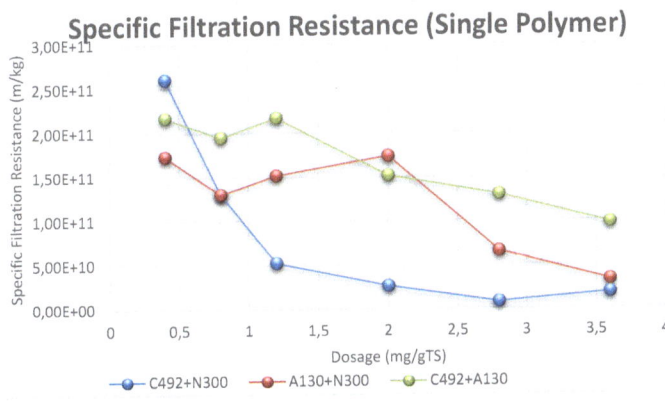

Figure 12: Specific Resistance value for C492+N300, A130+N300 and C492+A130

4.4 Comparison of the Specific Filtration Resistance

The specific filtration resistance result in a general context decreases as the dosage increases for all the considered flocculants. The result shows that the higher the dosage, the larger the flocs and the lesser required time to filter. Table 7 presents the specific filtration resistance value at optimum dosage for each of the flocculants used.

C492+N300 has the least range of resistance values followed by C492. Considering the floc structure, these two (C492+N300 and C492) produces larger floc structure compare to other flocculants and their filtration is faster. This agrees with the American Water Works Association (AWWA) that strong and small floc are the most preferable for filtration process. This will help to facilitate good filter performance and avert possibilities that can leads to premature floc breakthrough, and that too large flocs can easily clog filters (AWWA, 2000). A130 and C492+A130 has the highest range of resistance which confirms its relative difficulty to filter as it took longer time to filter during the experiment. This is related to the type of floc this flocculants formed with low permeability as it agrees with Bushell, et al., (2002) that with high permeability, large and porous flocs aid filtration.

Table 7: Specific filtration resistance at optimum dosage

Flocculants	Dosages (mg/gTS)	Filter Pressure (Bar)	Slope (s/m^6)	Filtration Resistance (m/kg)
C492	3.6	1.2	1×10^9	2.6034×10^{10}
A130	3.6	1.5	2×10^9	6.5085×10^{10}
C492+N300	2.8	1.2	3×10^8	7.8102×10^9
C492+A130	3.6	1.5	3×10^9	9.7628×10^{10}
A130+N300	3.6	1.5	1×10^9	3.2543×10^{10}

According to Ripperger, et al., (2012), the range of values of the filter resistance signifies the ease of filtration and it ranges between the values 10^8 (filtering very rapidly) to 10^{13} (nearly unfilterable). The result for all the flocculants considered therefore indicates that they are all within the easy to filter range.

Table 8: Correlation between SFR and assessed parameters

Parameter	C492	A130	C492+N300	A130+N300	C492+A130
SVI	-0.16031	-0.45747	0.38432	-0.62919	0.89575
Turbidity	0.27741	-0.40705	0.93861	0.34214	-0.93320

Comparing the physicochemical process result with the actual separation process, Table 8 presents a tabular figures for the correlation between the specific filtration resistance (SFR) values and the physicochemical optimization result (i.e. SVI and turbidity). Only the SVI for C492+A130 has a strong positive correlation, while the correlation with A130+N300 is negative. And turbidity for C492+N300 is very strong with SFR while C492+A130 has a strong negative correlation. This shows that result of the physicochemical optimization conditioning with this flocculants do not agree with the actual separation in most cases.

5. Conclusions and Further Studies

This study indicated the effectiveness of synthetic flocculants even at low dosage which also agrees with Brostow, et al., (2009). The cationic flocculants C492 (either as dual or single) shows that the higher the dosage, the larger the flocs and the more turbid the supernatant at dosages above the optimum. However, there is no positive significant effect in the SVI and turbidity of C492+N300 over C492. On the other hand, A130+N300 has lower SVI but more turbid supernatant compare to A130 under the similar range of dosages and the flocs structure of anionic flocculants A130 is not large, but small with high SVI.

With the actual separation process, dual-polymer C492+N300 gives the lowest specific filtration resistance of all the considered flocculants (due to its type of flocs) and this makes it the easiest to filter. Although all other flocculants as well have low specific filtration resistance, but anionic flocculants A130 (either as single or dual) took a longer time during the filtration process. Moreover, it could be deduced from the result that the timing it takes for pressure filtration to be completed do not implies that it has high specific filtration resistance just as established by Ripperger, et al., (2012). This is because there is no wide difference in their specific filtration resistance despite the time difference. In a haze, this study shows that choice of dual-polymer flocculation should be based on the properties of the flocculants which is highly dependent of the anticipated assessing parameters, as well as appropriate dosing concentration.

For further studies, it is recommend that more research should be done to verify if the polymer's combination ratio has an effect on the flocculation performance, and the efficiency of different polymer combinations under different ratio. Also, it will be of interest to know the effectiveness at higher dosage range for better understanding. And the flocs structure from each dosage is recommended for further investigation, which can as well contributes to research knowledge on its pelletization process.

References

AWWA. (2000). Operational control of coagulation and filtration processes. (C. Magin, Ed.) *Manual of water supply practices*, 1-33.

AWWA. (2011). *M37 Operational Control of Coagulation and Filtration Processes* (Third Edition ed.). America: American Water Works Association (AWWA). Retrieved November 2014

Britt, K. W. (1973). Retention of additives during sheet formation. *Tappi*, 83-86.

Brostow, W., Hagg Lobland, H. E., Sagar, P., & Singh, R. P. (2009). Polymeric Flocculants for wastewater and industrial effluent treatment. *Journal of Material Education, 31 (3-4)*, 157-166.

Bushell, G. C., Yan, Y. D., Woodfield, D., Raper, J., & Amal, R. (2002). On Techniques for the Measurement of the Mass Fractal Dimension of Aggregates. *Advances in Colloid and Interface Science, 95*, 1-50.

Chaiwong, N., & Nuntiya, A. (2008). Influence of pH, Electrolytes and Polymers on Flocculaation of Kaolin Particle. *Chiang Mai J. Sc., 35(1)*, 11-16.

Chen, P. J., Chang, S.-Y., Huang, J. Y., Bauman, E. R., & Hung, Y.-T. (2005). *Physicochemical Treatment Process (Chapter 13; Gravity Filtration)* (Vol. 3). (L. K. Wang, Y.-T. Hung, & N. K. Shammas, Eds.) Totowa, New Jersey: The Humana Press Inc. Retrieved from http://download.springer.com/static/pdf/400/bok%253A978-1-59259-820-5.pdf?auth66=1425283602_0116468b9bdf8d6f61cf9e4bed3689e2&ext=.pdf

Ebeling, J. M., Rishel, K. L., & Sibrell, P. L. (2005, February 2). Screening and Evacuation of polymers as flocculation aids for the treatment of aquacultural effluents. *Aquacultural Engineering, 33*, 235-249.

Fan, A., Turro, N. J., & Somasundaran, P. (2000). A study of dual Polymer flocculation. *Colloids and Surfaces: A physicochemical Engineering Aspects, 162*, 141-148.

Freese, S. D., Trollip, D. L., & Nozaic, D. J. (2003). *Manual for testing of water and wastewater treatment chemicals.* Water Research Commission. WRC. Retrieved November 2014, from http://www.wrc.org.za/Knowledge%20Hub%20Documents/Research%20Reports/1184%20web.pdf

Ghosh, R. (2006). *Principle of Bioseperation Engineering (Chapter 10; Filtration).* McMaster University, Canada: World Scientific Publishing Co. Pte. Ltd.

Girovich, M. J. (1996). *Biosolids Treatment and management; Process for beneficial use.* New York: CRC Press. Retrieved February 2015

Gregory, J., & Barany, S. (2011). Adsorption and flocculation by polymers and polymer mixtures. *Advances in colloid and Inteerfacce Science, 169*, 1-12.

Hande, K., Basaran, & Tasdemir, T. (2014). Determination of flocculation characteristics of natural stone powder suspension in the presence of different polymers. *Physicochemical roblems of Mineral Processing, 50(1)*, 169-184.

Hogg, R. (2000). Flocculation and dewatering. *Intenational Journal of Mineral Processing, V 58*, 223-236.

Lee, D.-J., Tay, J.-H., Hung, Y.-T., & He, P. J. (2005). *Physicochemical Treatment Processes (Chapter 17; Introduction to sludge treatment)* (Vol. 3). (L. K. Wang, Y.-T. Hung, & N. K. Shammas, Eds.) Totowa, New Jersey: The Humana Press Inc. Retrieved from http://download.springer.com/static/pdf/400/bok%253A978-1-59259-820-5.pdf?auth66=1425283602_0116468b9bdf8d6f61cf9e4bed3689e2&ext=.pdf

Mclaughlin, R. (2010, October 18). *NCSU.* Retrieved December 05, 2014, from http://www.ncsu.edu/wrri/pdfs/pastevents/esc102010/day2/McLaughlinPolymersChemistry.pdf

Moody, G., & Norman, P. (2005). *Solid-liquid Separation; Scale up of industrial equipment (Chapter 2; Chemical Pre-treatment).* West Yorkshire, UK.

Moss, N., & Dymond, B. (2013). Flocculation: Thoery and Aplication. *Ciba Speciality Chemistry*, 1-9. Retrieved January 2015, from http://www.siltstop.com/pdf/flocculation-theory_application.pdf

Mpofu, P., Addai-Mensah, J., & Ralston, J. (2003). nvestigation of the effect of polymer structure type on flocculation, rheology and dewatering behaviour of kaolinite dispersions. *Internatioal J. Mineral Processing, 71*, 247-268.

Ndum, C. (2013). *Filtration Resistance of the Conventional and Innovative Lignin Processing Methods.* Bachelor Thesis, Brandenburg Technical University Cottbus, Chair of Mineral Processing Technology, Cottbus, Germany.

Ou-Yang, H. D., & Santore, M. M. (1999). Thermodynamics and Kinetic Aspects of Bridging Flocculation. In K. Esumi (Ed.), *Polymer Interfaces and Emulsion* (pp. 269-310). New York, United State of America: Marcel Dekker, Inc.

Oyegbile, B., Ay, P., & Narra, S. (2015). Optimization of physicochemical process for pre-treatment of fine suspension by flocculation prior to dewatering. *Desalination and water treatment*, 1-11. doi:10.1080/19443994.2015.1043591

Pearse, M. J., Weir, S., Adkins, S. J., & Moody, G. M. (2001). Advances in mineral flocculation. *Mineral Enngineering, 14*, 1505-1511.

Razali, M. A., Ahmad, Z., & Ariffin, A. (2012, September 12). Treatment of pulp and paper mill wastewater with various molecular weight of polyDADMAC induced flocculation with Polyacrylamide in the hybrid system. *Advances in Chemical Engineering and Science, 2*, 490-503.

Ripperger, S., Gosele, W., & Alt, C. (2012). Filtration, 1. Fundamentals. *Wiley-VCH Verlag Gmbh & Co. KGaA, Weinheim, 14*, 677-709. doi:DOI: 10.1002/14356007.b02_10.pub2

Somasundaran, P., & Das, K. K. (1998, September 15-17). flocculation and selective flocculation; An overview. (S. Atak, G. Onal, & M. Celik, Eds.) 80-92.

Svarovsky, L. (2000). *Solid-Liquid Seperation* (Fourth Edition ed.). Oxford: Butterworth-Heinemann.

Wakeman, R., & Tarleton, S. (2005). *Solid-Liquid Seperation: Principle of industrial filtration.* UK: Elsevier.

Wu, M. R., & Theo, G. M. (2009, March 28). Flocculatio and rflocculation: Interplay between the adsorption behaviour of the components of a dual floocculant. *Colloids and Surfaces A: Physicochemical and Engineering Aspects, 341,* 40-45.

Yan, Y. D., Glover, S. M., Jameson, G. J., & Biggs, S. (2004). The flocculation efficiency of pilydisperse polymer flocculants. *Internatiional Journal of Mineral Processing, 73,* 161-175.

Yu, X., & Somasundaran, P. (1993, June 14). Enhanced flocculation with double flocculants. *Colloids and Surface A; Physicochemical and Engineering Aspects, 81,* 17-23.

Appendix

Pictorial view of the obtained floc with each of the flocculants (both as single and dual)

Single polymer

A130

C492

Dual Polymer

A130+N300

C492+N300

C492+A130

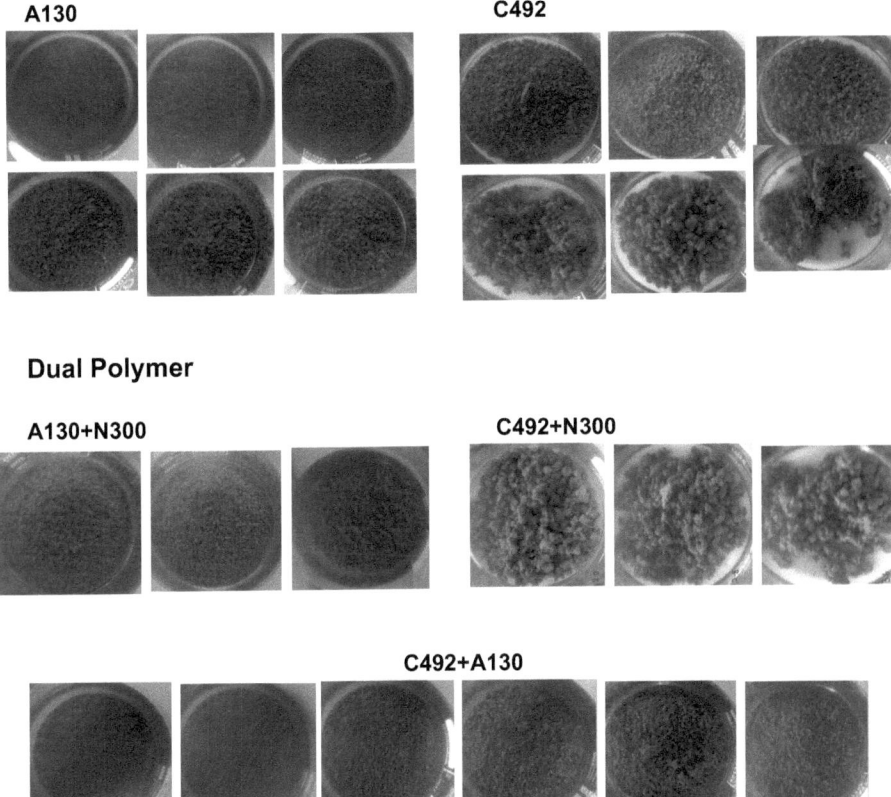